Healthy Body
Fighting Disease

by Kate Boehm Jerome

Table of Contents

Develop Language 2

CHAPTER 1 Healthy Body Systems 4
                Your Turn: Communicate 9
CHAPTER 2 Spreading Disease 10
                Your Turn: Summarize 15
CHAPTER 3 Preventing Disease 16
                Your Turn: Interpret Data 19

Career Explorations 20
Use Language to Give Examples 21
Science Around You 22
Key Words 23
Index 24

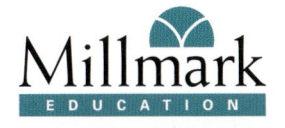

Millmark
EDUCATION

DEVELOP LANGUAGE

It is better to be healthy than sick! Even a cold can make you feel terrible.

The **germs** that spread **disease**, or illness, are too tiny to see. That is why you have to take special steps to protect yourself and others from them.

washing hands

Discuss the photos with questions like these.

How do you think some germs get into the air?

Germs get into the air by _____.

Why do you think it is important to wash your hands?

Washing your hands _____.

Why do you think the dentist is wearing a mask?

The mask _____.

What are other ways you can protect yourself from germs?

germs – tiny things that can make you sick

disease – a condition that causes illness

2 Healthy Body: Fighting Disease

dentist

sneeze

preparing food

Develop Language 3

CHAPTER 1

Healthy Body Systems

You feel good when your **body systems** are healthy and working well. Each body system is made up of a group of **organs**. Each organ does certain things. But each organ also works with other organs to allow the body system to do its main job.

For example, the teeth in your mouth break pieces of food into small pieces. These pieces move into your stomach. Liquids in your stomach break the pieces of food into smaller pieces. Both your mouth and your stomach help the **digestive system** do its main job of breaking down food.

Digestive System

- mouth
- stomach

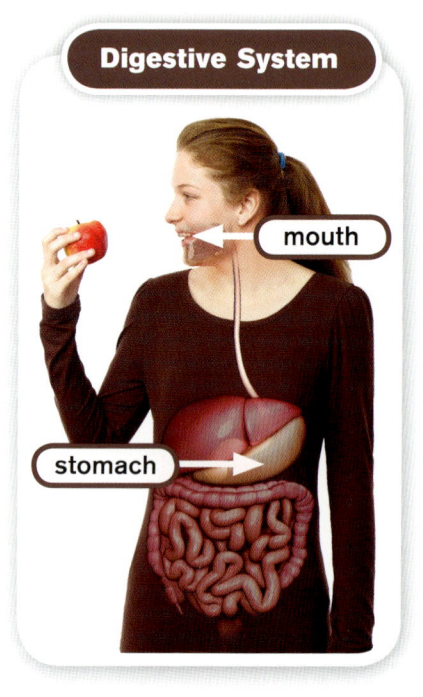

▲ The organs of the digestive system include the mouth and stomach. They help break down food so the body can get energy and other things it needs to stay healthy.

body systems – groups of organs that work together to do a job

organs – body parts that do things to help a body system work

digestive system – the body system that breaks down food

4 *Healthy Body: Fighting Disease*

You can see your mouth move. Sometimes you can feel food move to your stomach. But you cannot see or feel most of the activities in your body that keep you healthy.

That is because most of these activities happen in your **cells**. Cells are the smallest units of life. Most cells are too tiny to see. But microscopes make cells look bigger. That is how we know that groups of cells work together to form **tissues**. And groups of tissues work together to form your organs.

cells – the smallest units of life

tissues – groups of cells that work together to do a job

What Makes Up a Body System?

organs

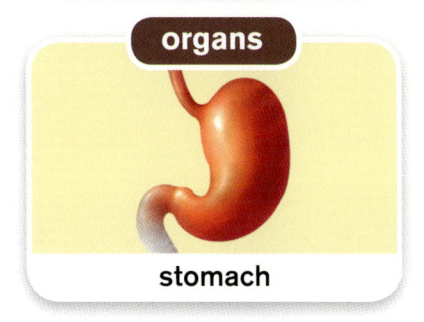

stomach

tissues

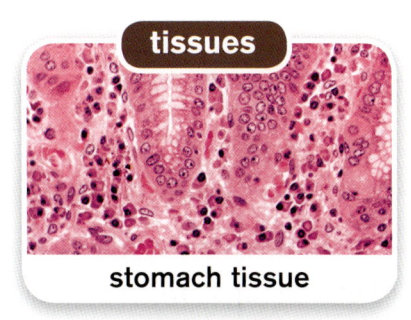

stomach tissue

cells

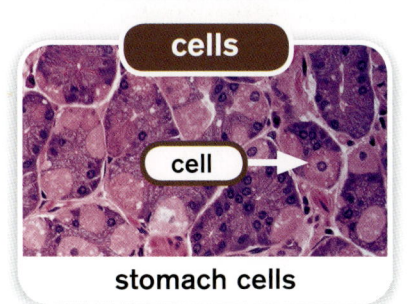

cell

stomach cells

Some Other Major Body Systems

System	Main Job
immune system	protects against disease
circulatory system	moves blood
respiratory system	allows you to breathe

KEY IDEA Groups of cells form tissues. Groups of tissues form organs. Groups of organs form body systems.

Chapter 1: Healthy Body Systems

Eating to Stay Healthy

Most cars need gasoline to run. In a similar way, you need food to keep your body systems working. Foods contain **nutrients**. Nutrients supply energy and other things your body needs. The main nutrients you need are listed in the chart on page 7. Water is one of the most important nutrients. In fact, the cells in your body are made mostly of water.

nutrients – substances the body needs to get energy or make the body work

6   *Healthy Body: Fighting Disease*

Different foods contain different nutrients. That is why you need to eat a variety of different foods every day. The healthiest foods contain the most nutrients. For example, fruits and vegetables contain many of the **vitamins** and **minerals** you need to eat each day.

vitamins – one of the groups of nutrients the body needs to be healthy

minerals – substances formed in nature; some are nutrients the body needs

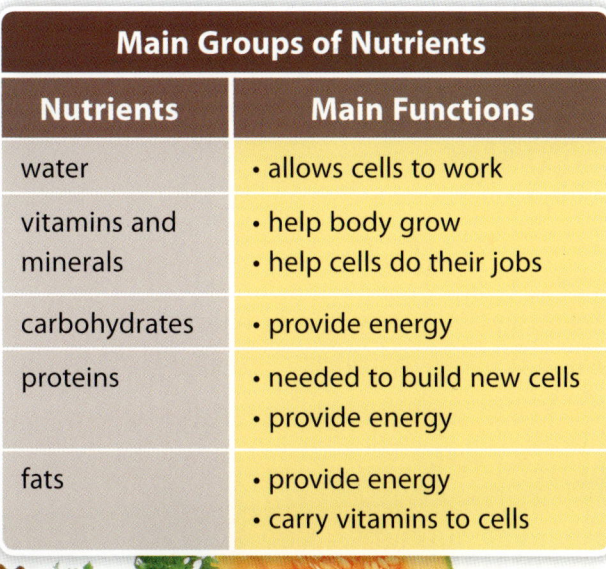

Main Groups of Nutrients

Nutrients	Main Functions
water	• allows cells to work
vitamins and minerals	• help body grow • help cells do their jobs
carbohydrates	• provide energy
proteins	• needed to build new cells • provide energy
fats	• provide energy • carry vitamins to cells

▼ Fruits and vegetables contain many different nutrients.

KEY IDEA Nutrients supply energy and other things your body needs. Different foods contain different nutrients.

Chapter 1: Healthy Body Systems

▲ This lunch has few nutrients and many calories.

▲ This lunch has many nutrients and fewer calories.

Calories

Besides nutrients, you also have to think about the **calories** in foods. Calories measure the amount of energy available in foods.

Everyone needs to eat a certain number of calories every day to stay healthy. However, if you eat more calories than you need, the extra energy is stored as fat in your body. Too much fat can be unhealthy.

Some foods have a high number of calories. If you use up your daily calorie amount on just a few high calorie foods, you probably will not get all the different nutrients you need.

calories – units that measure the amount of energy in foods

KEY IDEA Calories measure the amount of energy in foods.

8 *Healthy Body: Fighting Disease*

YOUR TURN

COMMUNICATE

Draw a picture of foods that you think would make up a healthy dinner.

With a friend, take turns asking and answering these questions.

1. Why do you think your dinner is healthy?
2. How could you make it healthier?

MAKE CONNECTIONS

Why do you think the digestive system is such an important body system?

USE THE LANGUAGE OF SCIENCE

What happens to the extra calories in foods that your body does not use?

Extra calories are stored as fat in your body.

Chapter 1: Healthy Body Systems 9

CHAPTER 2

Spreading Disease

Sometimes your body systems are not completely healthy. For example, **allergies** can make people cough and sneeze and feel very sick. But you cannot catch another person's allergies. That is because allergies are **noncommunicable diseases**. A noncommunicable disease cannot spread, or move, from one person to another.

However, there are other diseases that can spread. These illnesses are **communicable diseases**. When you get a cold or the flu, you have a communicable disease.

This person has allergies. But allergies cannot spread to other people.

allergies – reactions to certain substances
noncommunicable diseases – illnesses that cannot be spread
communicable diseases – illnesses that can be spread

10 *Healthy Body: Fighting Disease*

Communicable diseases are caused by **pathogens**, or germs. Pathogens include things like **bacteria** and **viruses**.

Most kinds of bacteria don't hurt us. They live in the water, air, and soil. They even live in and on your body. However, some bacteria cause disease. The bacteria in the picture cause strep throat.

Viruses are even tinier than bacteria. Viruses are not living things, but they can take control of cells. This damages the cells. It also allows the viruses to make more viruses. The viruses in the picture cause human flu.

bacteria

viruses

pathogens – germs that can cause disease
bacteria – tiny one-celled living things
viruses – tiny particles that can make more of themselves only inside living cells

KEY IDEA Communicable diseases are caused by pathogens such as bacteria and viruses.

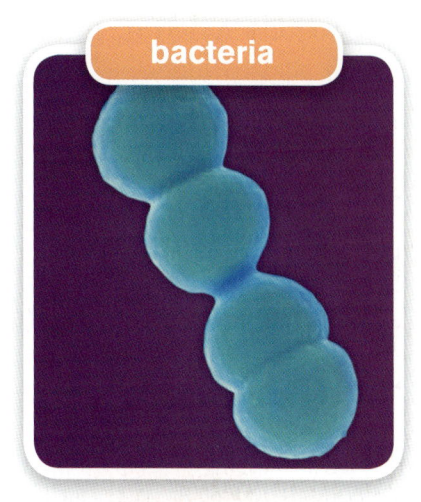

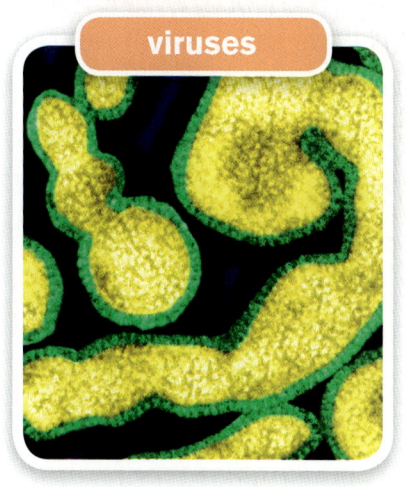

Explore Language

germen (Latin) = to start growing

One **germ** can grow into many **germs**. Plants **germinate** by growing a root and stem from a seed.

Chapter 2: Spreading Disease   11

How Disease Spreads

Pathogens move in lots of different ways. Some pathogens travel through the air. For example, viruses that cause a cold are sprayed into the air when you sneeze. If another person breathes in those viruses, that person might get sick with a cold.

▲ Cold viruses can spread through the air.

Those same cold viruses can also be passed in other ways. For example, if you do not wash your hands after you sneeze into them, the cold viruses remain on your hands. If another person touches your hands, the viruses can move to that person's hands. If that person then touches his mouth or nose, the viruses can enter his respiratory system and cause a cold.

SHARE IDEAS Explain why it is not a good idea to share a drinking glass with a sick person.

▶ Any object used by a person with a cold can carry cold viruses.

12 Healthy Body: Fighting Disease

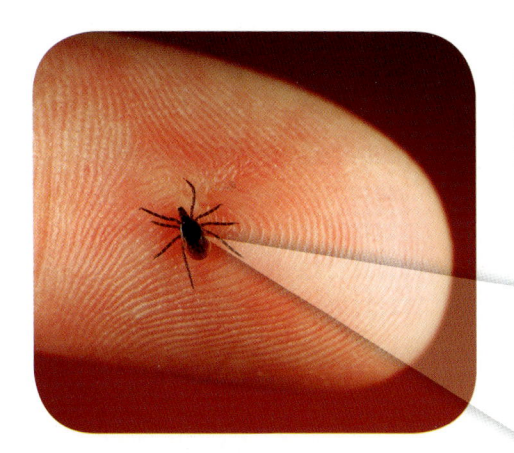

This tick can carry a pathogen that can cause Lyme disease in people.

Unfortunately, many pathogens can be found in water and foods. This means people can get sick from drinking water that contains pathogens. People can get sick from eating foods that contain pathogens.

Pathogens can also live in insects and other animals. People can get a disease if one of these animals bites them. For example, the tick in the photo can carry bacteria that cause Lyme disease. If the tick bites a person and breaks the skin, the bacteria can move into the person's blood. Then the person might get Lyme disease.

KEY IDEA
Communicable diseases can spread in many ways.

BY THE WAY...
Malaria is a communicable disease that is carried by one kind of mosquito. Between 300 and 500 million people get malaria each year around the world.

Chapter 2: Spreading Disease 13

Your Body's Defense

When pathogens enter your body, your **immune system** works to defend, or protect, you. For example, the strong liquids in your stomach destroy many pathogens that might be in the food you eat. Even the liquid in your mouth destroys some pathogens.

Special blood cells attack pathogens in your body. Some of these cells can make **antibodies**. Antibodies are substances that fight a particular pathogen. Once you develop antibodies against a certain pathogen, it will probably not be able to make you sick again.

▼ Your body forms antibodies against the pathogen that causes chicken pox. So once you have chicken pox, you are not likely to get it again.

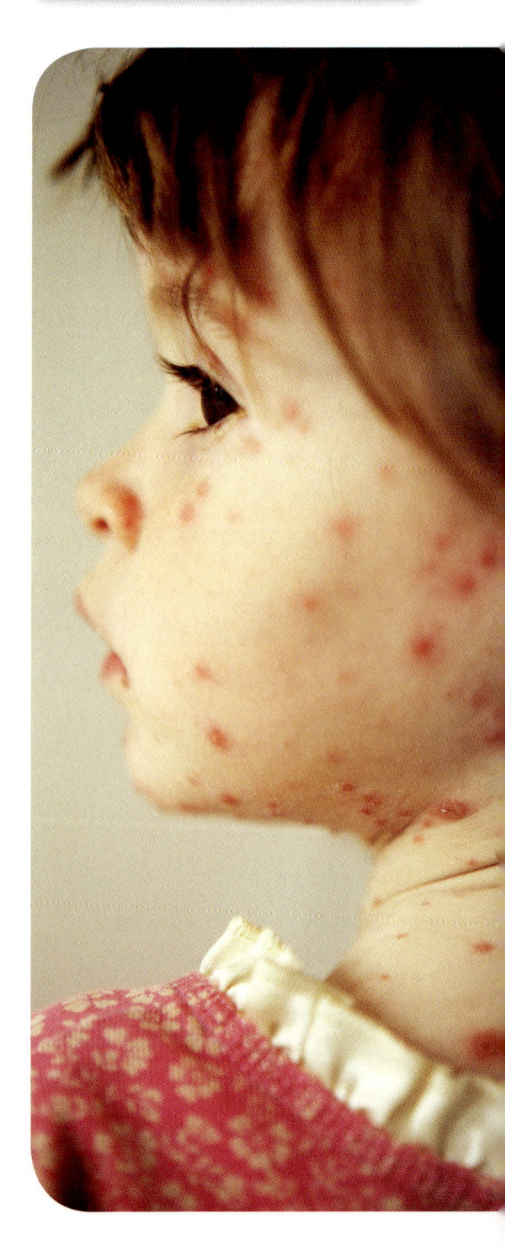

immune system – the body system that protects against disease
antibodies – substances made by certain blood cells that can destroy harmful pathogens

KEY IDEA Your body has ways to defend itself from pathogens.

14   *Healthy Body: Fighting Disease*

YOUR TURN

SUMMARIZE

Think about what you have learned about how pathogens spread. Make a chart like the one shown below. Fill in the chart with things you have learned.

Ways that Pathogens Can Spread	Example
1.	
2.	
3.	

MAKE CONNECTIONS

Why do you think it might be better to cover a sneeze with your elbow instead of your hand?

STRATEGY FOCUS

Monitor Comprehension

What did you find hard to understand in this chapter? What strategies did you use to understand it better?

Chapter 2: Spreading Disease 15

CHAPTER 3

Preventing Disease

The best way to stay healthy is to avoid getting sick. There are many ways you can prevent, or keep from, getting sick. Washing your hands is one of the most important ways to keep pathogens out of your body.

One way to avoid many serious diseases is to get a **vaccine** against the diseases. Vaccines make your body produce antibodies against some pathogens. If those pathogens enter your body in the future, you will not get the diseases they cause.

vaccine – a medicine that makes your body produce antibodies to protect against certain diseases

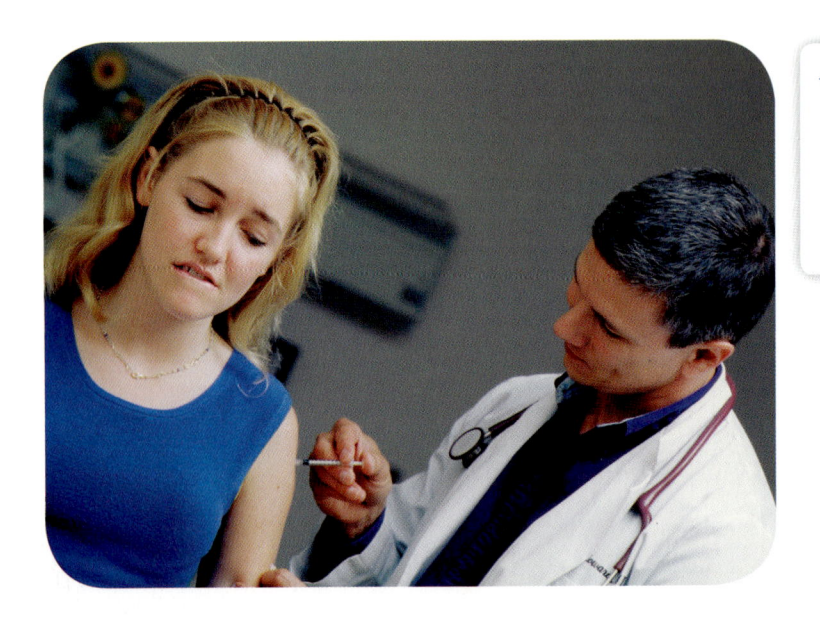

◀ This girl is getting a vaccine to prevent some diseases.

16 *Healthy Body: Fighting Disease*

Keeping your body systems healthy can also help you prevent disease. You already know that healthy food choices can help your body systems stay healthy. But you also need **exercise** to keep your body systems strong. Strong, healthy body systems are better able to fight disease.

Aerobic exercise makes your body use more oxygen. This helps your heart and lungs work better. Strength-building is another type of exercise. It helps build muscles.

The **food pyramid** is a useful guide to remind you what kinds of food to eat every day. It also reminds you to exercise every day.

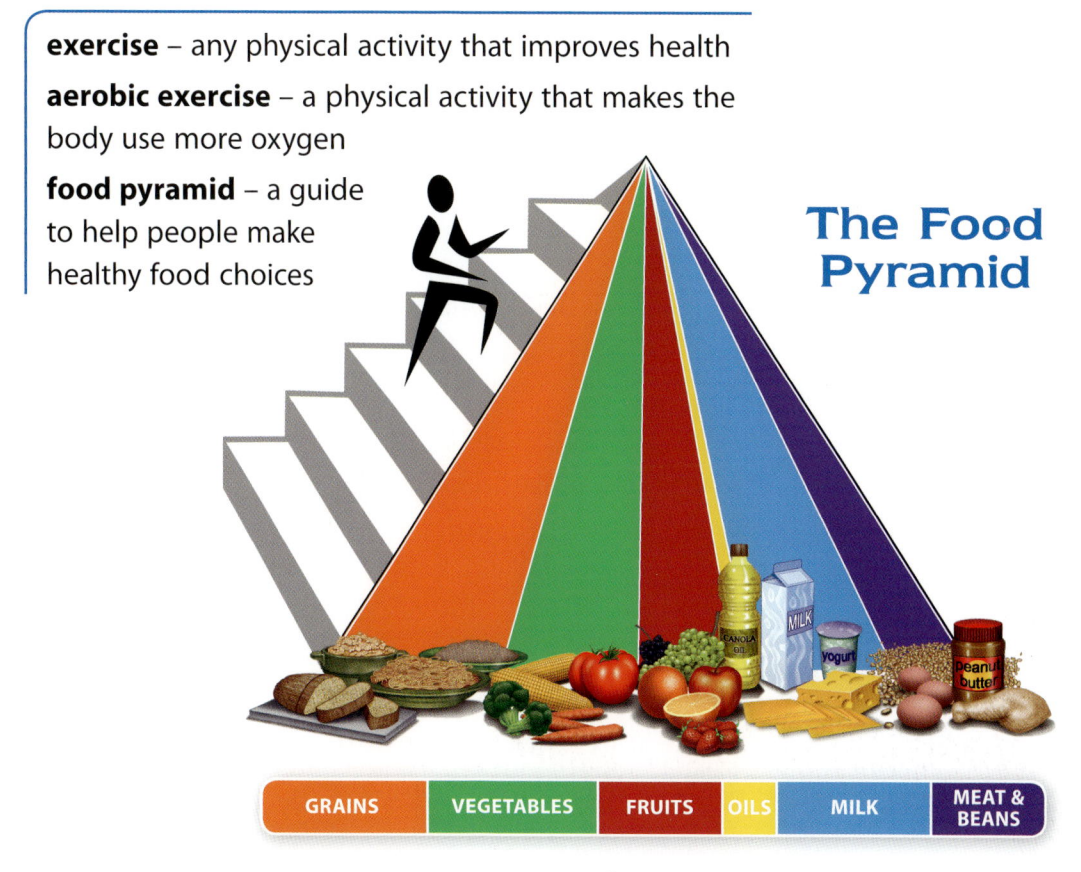

exercise – any physical activity that improves health

aerobic exercise – a physical activity that makes the body use more oxygen

food pyramid – a guide to help people make healthy food choices

The Food Pyramid

GRAINS | VEGETABLES | FRUITS | OILS | MILK | MEAT & BEANS

Chapter 3: Preventing Disease   17

Sleep helps your body grow and repair itself.

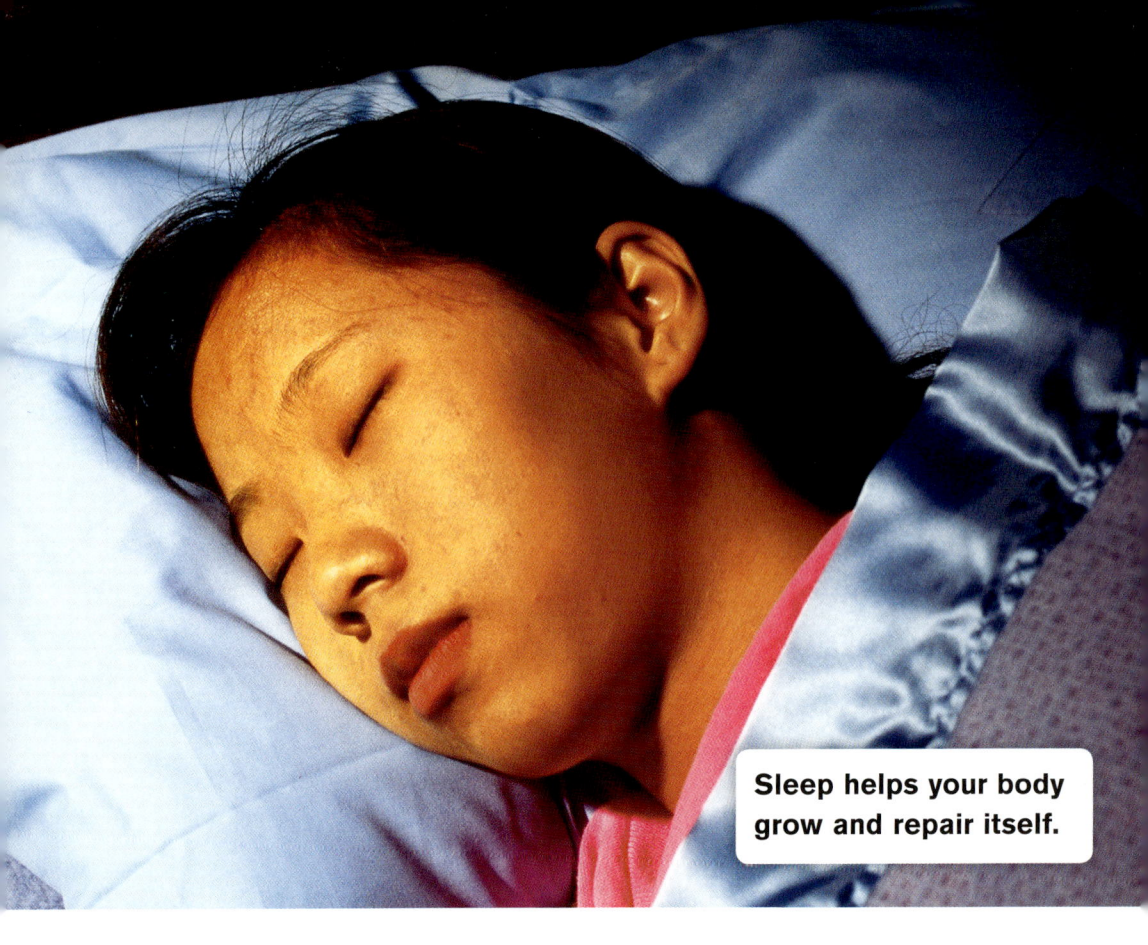

Sleep

Getting enough sleep is another thing you can do to prevent disease. Sleep allows your body to rest. Sleep also gives your body systems a chance to grow and repair themselves.

Sleep helps your brain work better. Scientists think that sleep may help you store information and remember things better.

KEY IDEA Activities such as eating healthy foods, exercising, and getting enough sleep can help keep your body healthy and prevent disease.

YOUR TURN

INTERPRET DATA

	Mon.	Tue.	Wed.	Thur.	Fri.
Hours of Sleep	8	6	7.5	9	8

The chart shows how many hours of sleep one student got each night of one school week. Use the chart to answer these questions.

1. Which night of the week did the student get the most sleep?

2. At this student's age, she needs at least 9 hours of sleep per night. How many nights did the student meet this need?

MAKE CONNECTIONS

Why do you think you need to do both strength-building exercise and aerobic exercise?

EXPAND VOCABULARY

Both the word **nutrient** and the word **nurse** come from the Latin word **nutrire**, which means "to nourish." How do nutrients and nurses nourish the body? Make a list of other words that come from **nutrire**. Share your list with a friend.

Chapter 3: Preventing Disease

CAREER EXPLORATIONS

Nursing People to Good Health

Do you like to take care of people? Are you interested in helping people stay healthy? Then maybe you would like to be a nurse.

Nurses work in many different places. They can work in doctors' offices. They also work in hospitals and clinics. Some nurses even travel around to visit people in their homes.

Nurses need at least two years of special training. Many nurses get a four-year college degree.

- Would you like to work as a nurse?
- Tell why or why not.

Healthy Body: Fighting Disease

USE LANGUAGE TO GIVE EXAMPLES

Examples in Words and Images

You can give examples to help someone understand an important idea. You can include examples in a written text, using **for example**, **such as**, and other phrases. You can also use pictures, captions, and charts to give examples.

EXAMPLES

Fruits **such as** pears and bananas have many nutrients.

Examples of Fruits
apples
oranges
peaches
plums

◀ Grapes **also** have many nutrients.

With a friend, look through this book. Discuss how the writer gives examples of important ideas. Look at the text, photos, captions, and charts.

Write with Examples

Choose a Key Idea in this book. Explain this Key Idea by giving examples.

- State the Key Idea in your own words.
- Give examples that provide more information about the Key Idea.
- Provide at least one picture or chart that gives examples of the Key Idea.

Words You Can Use

for example
such as
also
another example

SCIENCE AROUND YOU

5 steps to a healthy lifestyle

- Be active for an hour or more each day.
- Choose water as a drink.
- Eat more fruit and vegetables.
- Keep your vaccines up to date.
- Get a good night's sleep.

It is not that hard to take good care of your body! The five steps shown above can give you a good start to staying healthy and preventing disease.

- Sitting in front of the TV or computer for hours and hours is not in the chart. Why do you think that behavior can be unhealthy?

- Talk with a friend about how easy or hard it will be for you to do the five steps.

Healthy Body: Fighting Disease

Key Words

aerobic exercise (aerobic exercises) a physical activity that makes the body use more oxygen
Aerobic exercise strengthens the heart and lungs.

body system (body systems) a group of organs that work together to do a job
The digestive system is the body system that breaks down foods.

calorie (calories) a unit that measures the amount of energy in foods
Healthy foods often do not have as many calories as less healthy foods.

cell (cells) the smallest unit of life
Most cells are too tiny to see without a microscpe.

communicable disease (communicable diseases) an illness that can be spread
The common cold is a communicable disease.

disease (diseases) a condition that causes illness
Washing your hands can prevent the spread of disease.

food pyramid a guide to help people make healthy food choices every day
The food pyramid shows the different groups of foods people need.

noncommunicable disease (noncommunicable diseases) an illness that cannot be spread
An allergy is a noncommunicable disease.

nutrient (nutrients) a substance the body needs to get energy or make the body work
Water is an important nutrient that your body needs.

organ (organs) a body part that does things to help a body system work; a group of tissues that work together to do a job
The stomach is an organ in the digestive system.

pathogen (pathogens) a germ that can cause disease
Some bacteria and viruses are pathogens that cause disease.

tissue (tissues) a group of cells that work together to do a job
Stomach cells that work together form stomach tissue.

Index

aerobic exercise 17, 19
allergy 10
antibody 14, 16
bacteria 11, 13
body system 4–5, 9, 10, 17–18
calorie 8, 9
cell 5–7, 11
communicable disease 10–13
digestive system 4–5, 9
disease 2–3, 10–14, 16–18, 22

energy 4, 6–8
exercise 17–18, 19, 22
food pyramid 17
immune system 5, 14
microscope 5
mineral 7
noncommunicable disease 10

nutrient 6–8, 19, 21
organ 4–5
pathogen 11–14, 15, 16
sleep 18, 19, 22
tissue 5
vaccine 16, 22
virus 11–12
vitamin 7

MILLMARK EDUCATION CORPORATION
Ericka Markman, President and CEO; Karen Peratt, VP, Editorial Director; Lisa Bingen, VP, Marketing; David Willette, VP, Sales; Rachel L. Moir, VP, Operations and Production; Shelby Alinsky, Editor; Mary Ann Mortellaro, Science Editor; Kris Hanneman, Photo Research

PROGRAM AUTHORS
Mary Hawley; Program Author, Instructional Design
Kate Boehm Jerome; Program Author, Science

BOOK DESIGN Steve Curtis Design

CONTENT REVIEWER
Kefyn M. Catley, PhD, Western Carolina University, Cullowhee, NC

PROGRAM ADVISORS
Scott K. Baker, PhD, Pacific Institutes for Research, Eugene, OR
Carla C. Johnson, EdD, University of Toledo, Toledo, OH
Margit McGuire, PhD, Seattle University, Seattle, WA
Donna Ogle, EdD, National-Louis University, Chicago, IL
Betty Ansin Smallwood, PhD, Center for Applied Linguistics, Washington, DC
Gail Thompson, PhD, Claremont Graduate University, Claremont, CA
Emma Violand-Sánchez, EdD, Arlington Public Schools, Arlington, VA (retired)

TECHNOLOGY
Arleen Nakama, Project Manager
Audio CDs: Heartworks International, Inc.
CD-ROMs: Cannery Agency
ResourceLinks Website: Six Red Marbles

PHOTO CREDITS cover ©Eye of Science/Photo Researchers, Inc.; IFC and 15b ©David Safanda/iStockphoto.com; 1 ©Dr. Gary Gaugler/Photo Researchers, Inc.; 2 ©Tim Vernon/NHS Trust/Science Photo Library; 2-3 ©Lester V. Bergman/CORBIS; 3a ©Brand X Pictures/photolibrary; 3b ©Jim West/Alamy; 4 photo by Ken Karp for Millmark Education, illustration by Craig Bowman; 5a ©Jim Daugherty/Science Photo Library; 5b ©Dr. Gladden Willis/Visuals Unlimited/Getty Images; 5c ©Gladden Willis, M.D./Visuals Unlimited; 6 ©Creatas/photolibrary; 7 ©Julián Rovagnati/Shutterstock; 8a ©Vincent Giordano/Shutterstock; 8b ©Alamy; 8c ©objectsforall/Shutterstock; 9a ©Michelle D. Bridwell/PhotoEdit; 9b and 9c photos by Ken Karp for Millmark Education; 10, 15a, 20c, 24 ©David Young-Wolff/PhotoEdit; 11a ©Dennis Kunkel/Phototake; 11b ©Dr. Dennis Kunkel/Visuals Unlimited; 12a ©Jack Hollingsworth/Corbis; 12b ©Brand X/Corbis; 13a ©David M. Phillips/Photo Researchers, Inc.; 13b ©Scott Camazine/Photo Researchers, Inc.; 13c ©Sinclair Stammers/Photo Researchers, Inc.; 14 ©Ian Boddy/Photo Researchers, Inc ; 16 ©Will & Deni McIntyre/Photo Researchers, Inc.; 17 Illustration by Dino Idrizbegovic; 18 and 22e ©Doug Martin/Photo Researchers, Inc.; 20a ©Spencer Grant/PhotoEdit; 20b ©Robin Nelson/PhotoEdit; 21 ©Ievgeniia Tikhonova/Shutterstock; 22a ©Photodisc/Punchstock; 22b ©Eric Gevaert/Shutterstock; 22c ©carolgaranda/Shutterstock; 22d ©beerkoff/Shutterstock

Copyright ©2008 Millmark Education Corporation

All rights reserved. Reproduction of the whole or any part of the contents without written permission from the publisher is prohibited. Millmark Education and ConceptLinks are registered trademarks of Millmark Education Corporation.

Published by Millmark Education Corporation
PO Box 30239
Bethesda, MD 20824

ISBN-13: 978-1-4334-0247-0

Printed in the USA

10 9 8 7 6 5 4 3 2

Healthy Body: Fighting Disease